SAT®

Score-Raising
Math Dictionary
Second Edition

SAT®

Score-Raising
Math Dictionary

Second Edition

Jeanine Le Ny
with Colleen Schultz, Math Consultant

KAPLAN

PUBLISHING

New York

This publication is designed to provide accurate and authoritative information in regard to the subject matter covered. It is sold with the understanding that the publisher is not engaged in rendering legal, accounting, or other professional service. If legal advice or other expert assistance is required, the services of a competent professional should be sought.

Vice President and Publisher: Maureen McMahon
Editorial Director: Jennifer Farthing
Contributing Editors: Mark Ward, Brandon Jones, Katie Dewey
Development Editor: Eric Titner
Production Editor: Dominique Polfliet
Production Designer: Ivelisse Robles
Typesetter: Joe Budenholzer
Cover Designer: Carly Schnur

Published by Kaplan Publishing, a division of Kaplan, Inc.
1 Liberty Plaza, 24th Floor
New York, NY 10006

Printed in the United States of America

June 2008
10 9 8 7 6 5 4 3 2 1

ISBN-13: 978-1-4195-5287-8

Kaplan Publishing books are available at special quantity discounts to use for sales promotions, employee premiums, or educational purposes. Please email our Special Sales Department to order or for more information at kaplanpublishing@kaplan.com, or write to Kaplan Publishing, 1 Liberty Plaza, 24th Floor, New York, NY 10006.

How to Use This Book

Is math your most feared foreign language? Forget writing, vocabulary, and grammar…You can write an essay with the best of them, and Sentence Completion questions seem like a walk in the park. But does the Math Monster that makes up one-third of your SAT score keep you awake at night?

If so, you're definitely not alone. Math is a scary topic for tons of people, and that's why Kaplan came up with the *SAT Math Score-Raising Dictionary*. Kaplan wants to make sure that when you read math problems you really understand them. Translating the math language into English you can understand is the first step to mastery.

Using a cast of high school characters much like yourself, Kaplan will show you that it's possible to relate math to everyday life and to even (gasp!) make it enjoyable! Use this book as a tool for understanding those terms that come up again and again when you're studying for the math section of the SAT. Follow these steps when reading through the alphabetical list of words:

1. Carefully read the math term and its definition, even if you think you know what a word or phrase means. Read every part of the definition and pay close attention to its exact meaning—you may have been performing operations incorrectly because you have misunderstood what seems like a simple concept. This is the time to return to the basics and learn things right!

2. Read the sample sentence that follows the definition. The sentences use math words in the context of everyday life, detailing the lives of Vicky, Stephanie, Brad, Malcolm, and others at school, at cheerleading practice, in art class, and on weekend dates.

3. Try the sample problem that tests you on the concept. Answers to all problems begin on page 157.

Sit down, take a deep breath, and face that Math Monster with this dictionary by your side! With this book handy, your SAT math phobia should become a thing of the past.

Good luck!

KAPLAN

absolute value

the distance of a number from zero on the number line—uses the symbol ||; because absolute value is a distance, it is always positive

*Poor Malcolm only made himself feel worse when he used a number line to chart his dating life and came up with an **absolute value** of zero.*

Example:

The absolute value of 7 is 7; this is expressed |7| = 7.

The absolute value of −7 is also 7: |−7| = 7.

Every positive number is the absolute value of two numbers: itself and its opposite.

acute angle

an angle whose measure is between 0 and 90 degrees

*Ditzy Diana was a bit embarrassed when she found out that **acute angle** is not an angle that's good looking, but one that measures between 0 and 90 degrees.*

Which of the following measurements represents an acute angle?

(A) 60 degrees

(B) 90 degrees

(C) 100 degrees

(D) 180 degrees

add

to combine quantities to find a total

*Stephanie **added** up all of her extracurricular activities and concluded that 15 might be a bit too many to handle.*

Add 34 and 55.

additive inverse

the opposite of a number; a number and its additive inverse have a sum of 0

*The girls in the drama club rated Brad's dreamy smile a 10. Unfortunately his personality scored its **additive inverse**, negative 10.*

Which of the following is the additive inverse of $\frac{1}{2}$?

(A) $\frac{2}{1}$

(B) $-\frac{2}{1}$

(C) $-\frac{1}{2}$

(D) $\frac{1}{2}$

adjacent angles

two angles that share a common side and a common vertex

*Malcolm knew it was time to break from studying for the math quiz when he noted that the twins sitting back to back on the bleachers looked like perfect **adjacent angles**.*

In the following diagram, which of the following are adjacent angles?

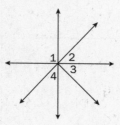

(A) angles 1 and 2

(B) angles 2 and 3

(C) angles 3 and 4

(D) angles 1 and 4

KAPLAN

alternate interior angles

two nonadjacent angles located between two lines, but on opposite sides of the transversal that crosses the two lines; if the lines are parallel, alternate interior angles are congruent (equal)

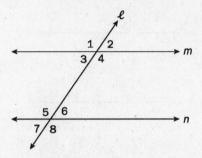

Angle pairs 3 and 6, and 4 and 5 are
alternate interior angles.

*Before track practice, Hannah and her teammate resembled **alternate interior angles** as they sat foot-to-foot and stretched toward each other.*

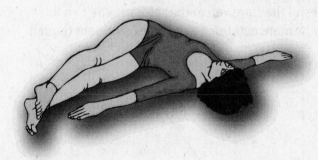

In the figure below, which angle pair can be classified as alternate interior angles?

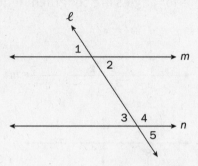

In the figure, line *m* is parallel to line *n*. If the measure of the angle 1 is 55 degrees, what is the measure of angle 8?

alternate exterior angles

two nonadjacent angles located on the outside of two lines, but on opposite sides of the transversal that crosses the two lines; if the lines are parallel, alternate exterior angles are congruent (equal)

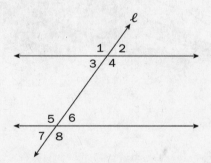

Angle pairs 1 and 8, and 2 and 7 are
alternate exterior angles.

*After track practice, Hannah and her teammate resembled **alternate exterior angles** as they sat back to back and stretched toward their feet.*

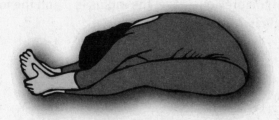

arc

a piece of a circle's circumference; if *n* is the degree measure of the arc's central angle, then the formula to find arc is $\dfrac{n}{360}(2\pi r)$

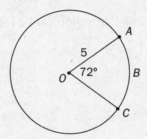

Vicki made her teammates on the cheerleading squad practice their backbends until each girl could perform a perfect arc.

In the figure on the previous page, the radius is 5 and the measure of the central angle is 72 degrees.

The arc length is $\dfrac{72}{360}$ or $\dfrac{1}{5}$ of the circumference:

$$\dfrac{72}{360}(2\pi)(5) = \dfrac{1}{5}(10\pi) = 2\pi$$

area

the number of square units in the interior of a region

*Diana could not pay attention in math class because she kept wondering how many bubbles it would take to fill the **area** of the room.*

What is the area of a rectangular region with a length of 6 *m* and a width of 4 *m*?

associative property

the property that states that the sum or product of terms remains the same even if the grouping is changed

$$a + (b + c) = (a + b) + c$$

$$a \times (b \times c) = (a \times b) \times c$$

*The **associative property** states that it doesn't matter if Brad's sister Bethany tags along on his date with her best friend April, OR if Brad tags along on Bethany and April's day at the mall. Three will always be a crowd.*

Which of the following is an example of the associative property of addition?

(A) $10 + (6 + 4) = (10 + 6) + (10 + 4)$

(B) $10(6 + 4) = (10 \times 6) + (10 \times 4)$

(C) $10 + (6 + 4) = (10 + 6) + 4$

(D) $(10 + 6) + 4 = 4 + (10 + 6)$

average

a single value that describes a set of numbers; it often represents the arithmetic mean

*Stephanie was amused when she realized that the **average** of all her test scores for the semester was the same number as her temperature: 98.6.*

What is the average (arithmetic mean) of 10, 12, 8, and 14?

axis

a vertical or horizontal number line drawn on a coordinate grid used to locate points on the grid; the horizontal line is the x-axis and the vertical line is the y-axis.

*Malcolm spotted Vicky heading down the football field to the bleachers and quickly made like a y-**axis** and crossed the field horizontally, hoping to "accidentally" bump into her as their paths met.*

The *x*- and *y*-axes are shown in the figure below.

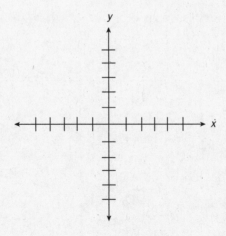

B

binomial

a polynomial with two terms

*Vicky thought it strange when Malcolm asked if she wanted to be the other half of his **binomial** instead of just asking her to be his girlfriend.*

Examples of binomials:

$3x + 2$

$7 - 3y$

$2x^3y - 2x$

$x^2 - 9$

bisector

a line, segment, or ray that divides an angle or segment into two equal parts

Malcolm's three-year-old brother was amazed when Malcolm cut a bisector into a square peanut-butter-and-jelly sandwich, transforming their dull lunch into two magical triangles.

If angle ∠*ABC* measures 68 degrees and is bisected by ray $\overrightarrow{BD}$, what is the measure of angle ∠*DBC*?

central angle

an angle whose sides are radii of a circle and whose vertex is at the center of the circle; the measure of a central angle is the same as the arc it intercepts

*The guys on the hockey team came up with a unique defensive formation that looks like a **central angle** in a ring.*

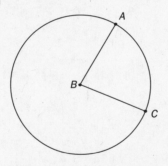

∠ABC is a central angle in circle B

In the figure above, if the measure of arc AC is 50 degrees, then what is the measure of central angle ∠ABC?

chance

the likelihood that an event will occur

*Hannah knew that only 10% of the girls at the starting line were faster than she, which meant that she had a 90% **chance** of winning the event!*

If the chance of rain is 30%, what is the chance it will not rain?

KAPLAN

chord

a segment contained within a circle whose endpoints are on the circle

*Malcolm always drew a **chord** through the O in his name, calling it his trademark signature.*

In the figure below, line segments $\overline{AB}$, $\overline{CD}$, and $\overline{EF}$ are each chords of circle *O*.

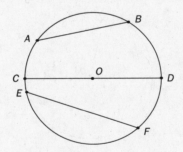

circle

the set of points equidistant from a center point

*Malcolm didn't get why was he always surrounded by a **circle** of laughing classmates while performing "The Robot" at school dances.*

The following figure is a circle with center point *O*.

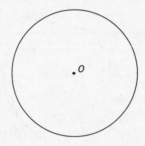

circumference

the distance around a circle; the formula for circumference is

$$C = \pi \times \text{diameter or } C = \pi d$$

*Diana often dreamed of joining the circus and training ferocious tigers to trot the **circumference** of the center circle under the big top.*

What is the circumference of a circle with a diameter of 10 *m*?

coefficient

the number directly preceding a variable; it is connected to the variable by implied multiplication

*Stephanie acted like a **coefficient** to her best friend Hannah—she was always right there in front of Hannah, multiplying her excitement in any situation.*

In the expression $4x^2 + 5x$, 4 and 5 are the coefficients.

combinations

the different groupings of a set of objects, without respect to order; the formula for combinations is $_nC_r = \dfrac{n!}{r!(n-r)!}$ where n is the total number of objects and r is the number of objects to be grouped

*The principal needed a **combination** of five students to be on a schoolwide cleaning committee, and no one seemed to want to volunteer!*

If there are five people to choose from for a committee of three, how many different combinations could be formed?

commutative property

the property that states that the sum or product of terms remains the same even if the order is changed

$$a + b = b + a$$

$$a \times b = b \times a$$

*Whether Hannah asks out Brad of if Brad asks out Hannah, the **commutative property** states that the result will be the same—they'll go to the movies on Saturday.*

Using the commutative property, what number belongs in the answer blank?

$$4 \times (8 + 12) = (8 + \underline{\quad}) \times 4$$

complementary

two angles whose measures sum to 90 degrees

*Stephanie couldn't believe how easy it was to identify **complementary** angles—two angles that add up to 90 degrees. If only it were that easy to find a complementary boyfriend!*

Two angles that measure 40 degrees and 50 degrees are complementary since 40 + 50 = 90.

congruent

figures that have the same shape and the same size—corresponding sides and corresponding angles have the same measure

*Brad spent an hour checking out the back of his head in a mirror. Was it really **congruent** to a jar like Hannah had said?*

If the two triangles below are congruent, what is the measure of angle *X*?

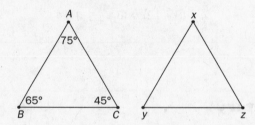

consecutive integers

integers in order that follow one after the other, for example, 5, 6, and 7; consecutive odd integers are numbers such as 1, 3, and 5 and consecutive even integers are numbers such as 10, 12, and 14

*Vicky, being the highly intelligent cheerleader that she is, often uses **consecutive integers** to get a crowd pumped. For example: "One, two, three, four! It won't be long before we score!"*

What are three consecutive integers whose sum is 57?

constant

a value that does not change

*Stephanie's score of 100 on every math test was a **constant** her class could expect to ruin the curve.*

> **Examples of constants:**
>
> In the expression x + 4, 4 is the constant term.
> In the expression 2 x 2 − 5x − 6, −6 is the constant term.

corresponding

parts of a figure that are in the same relative position

*Vicky was psyched when she found the **corresponding** megaphone-shaped earrings that matched her megaphone-shaped pendant.*

In the figure below, which two angles are corresponding?

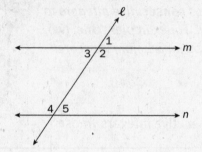

cosine (cos)

the ratio of the side adjacent to a given angle to the hypotenuse in a right triangle

*By calculating the **cosine** of the angle the javelin made with the ground at its highest point, Brad was able to see how much his distance increased over the track season.*

What is the cosine of angle *A* in the following right triangle?

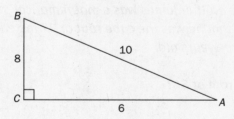

cubed

a number or expression that is used as a factor three times; an expression raised to an exponent of 3

*Sometimes Stephanie wished that she could **cube** herself so that she could be in three places at once.*

What is the value of 2 cubed?

cube root

a number that when used as a factor three times results in the radicand

*Malcolm knew his brother James was a mathematical genius when James told a lady that he was the **cube root** of 27 instead of simply saying he was three years old.*

What is the cube root of 64?

decimal

a value expressed using a decimal point and base ten place value

*Vicky likes to include **decimals** when talking about her height, which is exactly 5.06 feet tall.*

Examples of decimals:

    3.4
  110.24
    0.56
  −21.4

denominator

the number or expression in the bottom part of a fraction

*Vicky sure was glad that she never had to be the **denominator** of a two-girl cheerleading lift. Being on the bottom is hard work!*

In the fraction $\frac{10}{11}$, 11 is the denominator.

KAPLAN

diagonal

a line segment connecting any nonadjacent angles in a polygon

*The marching band formed a **diagonal** stretching from the bottom right corner to the top left corner of the football field.*

How many different diagonals can be drawn in a rectangle?

diameter

the line segment with endpoints on the circle that also passes through the center of the circle; the diameter is twice the length of the radius

*Diana surmised that the radius of the circular bald spot on the back of her math teacher's head is approximately 8 inches, making its **diameter** about 16 inches.*

If the radius of a circle is 6 m, what is the diameter of the circle?

difference

the result of subtraction

*The **difference** between the number of kids at Vicky's party—120—and the number her parents said she could invite—10—was 110.*

Find the difference between 68 and 33.

dilation

an image of an object created by enlarging or reducing the size and distance of the object; the symbol D_2 denotes a dilation scale factor of 2 (the image will be twice as large as the original figure and twice as far away from the origin)

*Malcolm wished he could give Vicky's smile a **dilation** scale factor of negative 2. Sure, it would be twice as small, but also twice as close to him. Is that so wrong?*

After a dilation of scale factor 3, what is the image of the point (4, −5)?

direct variation

a relationship in which the variable y changes directly (at the same constant rate) as x does; often expressed as $y = kx$, where k is a nonzero constant

*Brad soon realized that his poor behavior in class had **direct variation** effect to the time he spent in detention.*

Examples:

If a unit of Currency A is worth two units of Currency B, then A = 2B. If the number of units of B were to double, the number of units of A would double, and so on for halving, tripling, etc.

distance formula

the formula used to find the length of a line segment; the distance between two points (x_1, y_1) and (x_2, y_2); the formula is

$$d = \sqrt{(x_1 - x_2)^2 + (y_1 - y_2)^2}$$

*As Hannah watched her opponent take her final long jump, Hannah used the **distance formula** to figure out how far she'd have to jump to beat her.*

What is the distance between the points (3, 2) and (−1, −1)?

distributive property

the property that states that the product of a sum (or difference) of two terms is the same as the sum (or difference) of their products; this property states that multiplication distributes over addition and subtraction

$$a(b + c) = ab + ac$$

*Hannah and Vicky both needed dates to the spring dance, so Malcolm decided to use the **distributive property** and go with both of them— spending equal time with each. What a guy!*

Using the distributive property, what number belongs in the answer blank?

$$4(20 + 5) = (4 \times 20) + (\underline{\ \ } \times 5)$$

divide

to split into equal groups; the opposite of multiplying

*Stephanie insisted that she and her date go "Dutch," so they **divided** the dinner bill by two.*

Divide 54 by 9.

dividend

the value being divided in a division problem

*Vicky's mom gave her $30. Unfortunately it was a **dividend** to be split between her and her brother.*

In the statement 33 ÷ 3 = 11, what is the dividend?

divisor

the value being divided by in a division problem

*Vicky was glad that the $30 dividend had a **divisor** of only 2. As it was, she could barely buy a movie ticket with her share.*

In the statement 63 ÷ 7 = 9, what is the divisor?

domain

the set of values for which a function is defined; also known as the input values for a function

*The **domain** for Diana's personality included kind, sweet, generous, and giving. Selfish and cruel were definitely not in the **domain**.*

What is the domain of the function $f(x) = \dfrac{1}{1 - x^2}$?

edge

a line segment of a three-dimensional solid where two faces meet

*Stephanie's toes hung over the **edge** of the platform as she prepared to dive into the pool.*

In the diagram below, what line segment forms the edge joining sides 1 and 2?

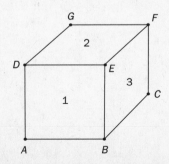

equation

a mathematical sentence that states that two expressions are equal

*Although they were different, they were like two sides of an **equation**.*

Solve the equation $3x - 4 = -16$ for x.

KAPLAN

equilateral triangle

a triangle in which all three sides are equal; the angles are also equal and measure 60 degrees each

*Just because Vicky suggested cutting two cheerleaders from the stunt so they could form an **equilateral triangle** with their pyramid doesn't mean she's obsessed with perfection. Does it?*

In the figure below, what is the measure of side $\overline{YZ}$?

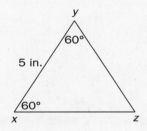

equivalent

having the same value

*Stephanie didn't want to be rude, so she hoped that eating tons of garlic and onions and more garlic was the **equivalent** of saying out loud that she did not want her date to kiss her.*

Which of the following fractions is not equivalent to $\frac{6}{8}$?

(A) $\frac{3}{4}$

(B) $\frac{9}{12}$

(C) $\frac{12}{16}$

(D) $\frac{16}{20}$

exponent

a number that tells how many times the base is used as a factor

*Diana was not amused when the math teacher added an **exponent** of 4 to the 2 pages of homework he'd assigned for the weekend, giving the class 16 pages to complete on their days off!*

In the expression $3 \times 5^4 + 8$, what number is an exponent?

(A) 3

(B) 5

(C) 4

(D) 8

exponential growth

a sequence or pattern that continues based on the influence of an exponent or exponents

*Mr. Martin was serious when he told the class that their homework would have an **exponential growth** of two each time someone spoke out of turn.*

Example:

The pattern 1, 2, 4, 8, 16, 32,... is an example of an exponential growth pattern. Each term in the sequence is an increasing power of 2:

$2^0 = 1$

$2^1 = 2$

$2^2 = 4...$

expression

a mathematical phrase that contains numbers, variables, and operations; an expression does not contain an equal sign or an inequality symbol

When Stephanie tried to explain to Diana that $x^2 + 2x - 3$ is an **expression,** *Diana disagreed and said that "it's raining cats and dogs" is an expression.*

Which of the following is an example of an expression?

(A) $x^2 + 2x - 3$

(B) $3x - 1 = 10$

(C) $x \leq -4$

(D) $4x^3 + 9 = 5$

exterior angle

an angle on the outside of a polygon created by extending a side of the polygon

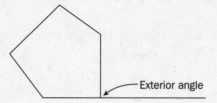

Exterior angle

*Hiding within the **exterior angle** of the open door, the students eavesdropped on the teachers arguing in the classroom.*

What is the measure of an exterior angle of a regular pentagon?

face

a polygon that is part of the surface of a three-dimensional solid

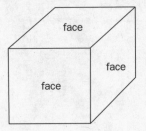

*Malcolm left a sweaty palm print on the heart-shaped **face** of the red velvet box he handed to Vicky on Valentine's Day.*

What is the area of the face of a rectangular prism with dimensions 6 cm and 10 cm?

factor

a positive integer that divides into a number without a remainder

*Vicky decided to get a library card when she found out that the number of books she had read that year, two, was merely a **factor** of 126, the number Stephanie had read.*

For example, 6 is a factor of 12 and 5 is a factor of 10.

factorial

the product of that whole number and each of the natural numbers less than the number; it is written as $n! = n \times (n-1) \times (n-2) \times \ldots \times 1$

*Finally Diana found out a real-life use for algebra when she figured out, by using **factorial**, the total number of color combinations she could use in her fashion-design class.*

If five different books are placed on a shelf, how many different arrangements are there for the five books?

FOIL

an acronym to recall the steps for multiplying two binomials: F (multiply the first terms in each binomial), O (multiply the outer terms), I (multiply the inner terms), L (multiply the last terms) in each binomial

*Stephanie knew she had to quit being Diana's math tutor when she told Diana to use **FOIL** to multiply binomials and the girl went into the kitchen to get the aluminum foil.*

Multiply the binomials $(x + 5)$ and $(x - 4)$ using FOIL.

function

a mathematical relationship or equation where every member of the domain (x-value) corresponds with exactly one member of the range (y-value)

*Hannah's allergic reactions to nuts acted like a **function**—for every different nut she put in her body, a different symptom resulted.*

Which of the tables below does not represent a function?

(A)

x	y
1	5
2	6
3	7
4	8

(B)

x	y
0	10
2	8
4	6

(C)

x	y
1	3
2	5
3	7
1	9

(D)

x	y
−1	0
0	2
1	4
2	0

greater than

symbol (>) that shows the first term has a larger value than the second term

*Malcolm's allowance is **greater than** his little brother's.*

Which of the following is true?

(A) 4 > 5

(B) 12 > 14

(C) 8 > 7

(D) 1 > 2

greater than or equal to

symbol (≥) that shows the first term has an equal or larger value than the second term

*Brad figures that if a girl has in her wallet a **greater than or equal** to the amount of cash he has, then she should pay for the date.*

Solve $x + 4 \geq 6$ for x.

greatest common factor (GCF)

the largest number that is a factor of two or more integers

*The **greatest common factor** between the cheerleaders is that they are all blondes.*

Find the GCF of 36 and 48.

height

the perpendicular distance between a vertex and the opposite side of a polygon

*Hannah's **height** gives her an advantage over the shorter girls competing in the high jump.*

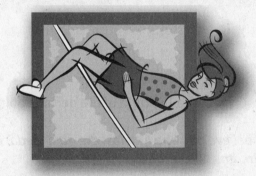

The height of each of the triangles below is labeled *h*.

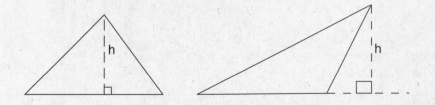

hundreds

the place value three spaces to the left of the decimal point

*Malcolm hoped and prayed that his weight would one day enter into the **hundreds**.*

In the decimal number 1,608.532, the number _____ appears in the hundreds place.

hundredths

the place value two spaces to the right of the decimal point

*Hannah couldn't believe that she had lost the gold medal in the hurdles race by two **hundredths** of a second!*

Each of the numbers 0.37, 5.26, and −90.18 are expressed to the hundredths place, or two places to the right of the decimal. The digits 7, 6, and 8 are each located in the hundredths place.

hypotenuse

the longest side of a right triangle; the side opposite the right angle in a right triangle

*Instead of walking the length of the soccer field and then the width, Vicky cut across diagonally by taking the **hypotenuse** route and saved herself a few minutes.*

In the right triangle below, side $\overline{AC}$ is the hypotenuse.

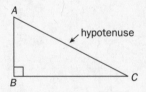

identity

the number when added or multiplied by a given number gives a result of the same given number; the additive identity is 0 and the multiplicative identity is 1

$$a + 0 = a$$

$$a \times 1 = a$$

*When Brad whined about not getting the part he wanted in the school play, the director told Brad to act his age, not the **multiplicative identity**.*

Which of the following illustrates the additive identity property?

(A) $16 + -16 = 0$

(B) $-16 + 0 = -16$

(C) $16 \times \dfrac{1}{16} = 1$

(D) $16 \times 1 = 16$

image

the result of a transformation of an object

*Diana smiled at her **image** in the mirror to see if she had lipstick on her teeth.*

Which of the following represents the image of triangle *XYZ* after a reflection over the *x*-axis?

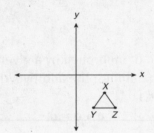

(A)   (B)

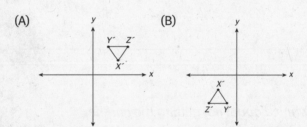

(C)   (D)

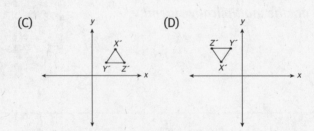

imaginary number

the square root of a negative number; is written in the form xi where $i = \sqrt{-1}$

*Diana wondered if she could somehow get an **imaginary** friend to do her homework on imaginary numbers.*

Examples of imaginary numbers:

$$\sqrt{-9} = 3\,i$$
$$\sqrt{-25} = 5\,i$$
$$\sqrt{-36} = 6\,i$$

---

improper fraction

a fraction with a numerator that is larger than the denominator

*Brad is so conceited that if he were a fraction he'd be an **improper fraction**—his head is way bigger than the bottom half of his body.*

The fractions $\frac{12}{11}$, $\frac{20}{9}$, and $\frac{101}{95}$ are in improper fraction form.

inequality

a mathematical statement that uses one of the symbols $>$, $<$, $\geq$, $\leq$, or $\neq$ to state a relationship between expressions

*Hannah fumed at the **inequality** of her school's sports program: the boys receive greater funding and better equipment than the girls.*

Which of the following does not represent an inequality?

(A) $3 + 4 \neq 8$

(B) $6 > 2 + 3$

(C) $14 - 3 \leq 11$

(D) $4 \div 2 = 2$

inscribed angle

an angle whose sides are chords of a circle and whose vertex is on the circle; the measure of an inscribed angle is half of the measure of the arc it intercepts

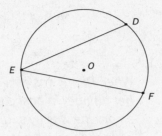

∠DEF is an inscribed angle in circle O

*Instead of slicing it normally, Malcolm **inscribed** two **angles** into the pepperoni pizza, with the point at the crust. Then he removed one and ate it. It still tasted delicious.*

What is the measure of inscribed angle ∠DEF if the measure of the intercepted arc DF is 96°?

integers

the set of whole numbers and their opposites; they include negative whole numbers and 0

*When Diana was asked to pick an **integer** for her number in basketball, she proved how smart she was by suggesting −3. The coach changed the rules to positive integers.*

Examples of integers:

−15

−4

0

1

457

in terms of

to isolate the one variable on one side of the equation, leaving an expression containing the other variable on the other side of the equation

*Cheerleader Vicky often describes herself **in terms of** peppiness.*

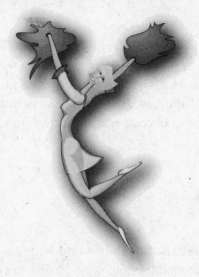

Solve the equation $3x - 10y = -5x + 6y$ for x in terms of y.

intersection

the set of elements common to both Set A and Set B; can be expressed as $A \cap B$

*Brad and Stephanie were the **intersection** of the artsy kids and the jocks—both were soccer players who could also paint and draw.*

For example, if Set A = {1, 2, 3} and Set B = {3, 4, 5}, then $A \cap B$ = {3}.

inverse variation

a relationship where as one variable increases, the other variable decreases at the same constant rate; can be expressed as $xy = k$, where x and y are variables and k is a constant

*Brad finally understood the theory of **inverse variation** when he had to walk home from school. It usually took 10 minutes by car but on foot it took longer because he was traveling slower.*

A commonly used inverse relationship is **rate × time = distance**, where distance is constant. For example, imagine having to cover a distance of 24 miles. If you were to travel at 12 miles per hour, you'd need two hours. But if you were to travel at half your rate, you would have to double your time. This is just another way of saying that rate and time vary inversely.

irrational numbers

real numbers with a location on the number line that cannot be expressed exactly as a fraction or decimal; irrational numbers are nonrepeating, nonterminating decimals

Sometimes Vicky talked too much, going on and on and on like an **irrational number** *with nonterminating decimals.*

The following are irrational numbers:

$$\sqrt{2},\ \sqrt{3},\ 0.10110111...,\ \text{and}\ \pi.$$

isolate

the process of solving for a variable; to get the variable by itself

In order to study, Stephanie needed to **isolate** *herself from the world. Otherwise she'd be too tempted to procrastinate.*

In the following equation, isolate the variable y.

$$2x + 3y = 9$$

isosceles

a triangle or trapezoid that has two equal sides; the base angles of the figure are also equal

For fun Malcolm and his buds bet on who could draw an accurate isosceles triangle the fastest...without a ruler.

Which of the following sets could represent the measures of the sides of an isosceles triangle?

(A) {1, 2, 3}

(B) {3, 5, 7}

(C) {5, 8, 8}

(D) {7, 4, 5}

least common denominator

the smallest common multiple of two or more denominators

*Stephanie wished that finding world peace was as simple as finding the **least common denominator** in fractions. Once you figure out the common multiple, anything is possible.*

Find the least common denominator for the fractions $\frac{3}{5}$ and $\frac{5}{7}$.

least common multiple (LCM)

the smallest multiple of two or more integers

*Finding a **least common multiple** is kind of like shopping for jeans—you have to use trial and error until you get a good fit.*

Find the LCM of 12 and 15.

leg

one of the two shorter sides of a right triangle; the two legs form the right angle of any right triangle

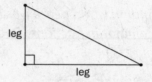

*Vicky fiddled with the bottom **leg** of her triangle-shaped earring as she gossiped on the phone with her best friend.*

True or False:

If the measure of a leg and the hypotenuse of a right triangle are 6 and 10, respectively, then the measure of the other leg is 7.

less than

symbol (**<**) that shows the first term has a smaller value than the second term

*Brad thought his performance in the talent show was superb yet the applause was **less than** he expected.*

Which values from the set {0, 5, 11} satisfy the inequality 10 < *x*?

less than or equal to

symbol (≤) that shows the first term has a value equal to or smaller than the second term

*Stephanie needed to find a gift for her friend that cost **less than or equal to** 20 bucks, because that was all she had in her wallet.*

If 2*x* ≤ 20, then what is the solution set for *x*?

like terms

algebraic terms with the exact same variables and exponents

*Hannah doesn't care for surprises—or multiple variables. That's why she only hangs out with like-minded people and enjoys solving equations containing **like terms**.*

Examples of like terms:

$3x$ and $4x$
$6a^2$ and $-10a^2$
$-2xy$ and $5xy$

line

an infinite set of points contained in a straight path in two opposite directions; line AB can also be written as $\overleftrightarrow{AB}$, $\overleftrightarrow{BA}$, or by a single lowercase letter that labels the line

*Vicky watched in amazement as the school's enormous marching band advanced onto the football field and stretched into a **line** that seemed to go on forever.*

What is the name of the line in the figure?

line segment

the infinite set of points that forms a straight path between two endpoints; line segment
AB can also be written as $\overline{AB}$ or $\overline{BA}$

*Malcolm took his job as hall monitor seriously and made the six offenders without passes form a **line segment** outside the principal's office.*

True or False:

The order of the points does not matter when naming a line segment.

linear equation

an equation of degree 1; the standard form is $y = mx + b$

*Vicky plugged her recent purchases into a **linear equation** that, when graphed, made a line that showed the more money she spent, the higher her debts grew.*

Which of the following represents a linear equation?

(A) $y = 4x + 4$

(B) $y = 3x^2$

(C) $y = x^3 + 1$

(D) $y = \dfrac{2}{x} + 1$

lowest terms

the simplest form of a fraction or expression

*Oh, how Malcolm wished he could simplify the complicated school hierarchy to its **lowest terms**.*

Reduce the fraction $\dfrac{28}{36}$ to lowest terms.

KAPLAN

mean

a statistical measure that is the result when the sum of a set of numbers is divided by the number of values in the set

*Vicky freaked out when she received a 65 on her English exam. The poor score sent her previous **mean** from 95 all the way down to 85.*

Shira received scores of 84, 90, and 92 on her first three tests. What is her mean score?

median

the value that falls in the middle of the set when the numbers are written in order; if there is an even number of values in a set, take the average of the two middle numbers

*Brad received scores of 6.4, 6.6, 6.8, 7.2, and 7.8 on his performance at the talent show, giving him a **median** score of 6.8.*

Steven's five test scores are 88, 86, 57, 94, and 73. What is his median grade?

midpoint

the location halfway between two given points; the formula is

$$\left(\frac{x_1 + x_2}{2}, \frac{y_1 + y_2}{2} \right)$$, where (x_1, y_1) and (x_2, y_2) represent the two

endpoints (the average of the x-coordinates of the endpoints and the average of the y-coordinates of the endpoints)

*Diana stopped at the **midpoint** between her house and the school when she realized that it was Saturday and she could have slept in.*

What is the midpoint of the line segment with endpoints (3, 5) and (9, 1)?

KAPLAN

minus

the symbol for subtraction (–), or the act of subtracting

*Stephanie considered Brad the most perfect guy at school...**minus** the arrogant personality, of course.*

What is 14 minus 8?

mixed number

a number expressed as an integer followed by a proper fraction

*Brad bragged to his date that he had eaten a **mixed number** of pancakes for breakfast. Stephanie assumed he had eaten a few different kinds, but then Brad explained that he had eaten $6\frac{1}{2}$ in total.*

The values $6\frac{1}{4}$, $-20\frac{5}{6}$, and 15 are all in mixed number form.

mode

the value that appears most often in a set; if there are two values that appear most often, the set is bimodal and if no value appears most often the set has no mode

*As an experiment, Vicky recorded her moods throughout day and found the **mode** of her moods was perpetually peppy.*

Charlotte's quiz grades were 88, 57, 68, 85, 99, 93, 93, 84, and 81. What is the mode of her scores?

monomial

a polynomial with one term

*If Malcolm didn't want to be the only **monomial** at the Spring Fling, he had to get a date quick!*

Which of the following is not an example of a monomial?

(A) $5x^3y^2$

(B) $x + 4$

(C) mn

(D) $12z$

multiple

a positive integer that a whole number divides into without a remainder

*Vicky and her three friends each had five dollars, so it worked out perfectly that the check at the burger joint cost $20, a **multiple** of five.*

Example:

24 is a multiple of 12 and 100 is a multiple of 20.

multiplicative inverse

the reciprocal of a number; a number and its multiplicative inverse have a product of 1

*One might describe Hannah and her new boyfriend, George, as **multiplicative inverses**. They may seem like total opposites, but they really complement each other to make a perfect unit.*

Which of the following is the multiplicative inverse of −4?

(A) $-\dfrac{1}{2}$

(B) $-\dfrac{1}{4}$

(C) $\dfrac{1}{4}$

(D) 4

multiply

the act of combining groups of equal amounts, or repeated addition; the opposite of division

*Malcolm didn't know how to explain to his little brother why there were suddenly four bunnies instead of two in Fluffy and Puffy's cage, so he simply said that bunnies can **multiply**.*

Multiply 6 times 3.

natural numbers

the set of counting numbers (1, 2, 3, 4, 5,...)

Vicky used only natural numbers in her cheers, because there was no place for fractions, decimals, negative numbers, or zeros at a pep rally.

Which of the following is an example of a natural number?

(A) 0

(B) $6\frac{1}{2}$

(C) 65

(D) 10.2

negative

a value that is less than zero, or located to the left of zero on a number line

*Unfortunately on a scale of 1 to 10, the girls rate Malcolm's pale scrawny body in the **negative** numbers.*

> **Examples of negative numbers:**
>
> $-1, -3, -4.5, -\dfrac{1}{2}$ and -601

numerator

the number or expression in the top part of a fraction

*Malcolm thought it was so cute when his brother called the top bunk his **numerator** bed.*

In the fraction $\dfrac{3}{4}$, 3 is the numerator.

obtuse angle

an angle whose measure is between 90 and 180 degrees

*Hanna had to pull her locker open to **an obtuse angle** in order to get her stuff out.*

Examples of obtuse angles:

91°

115°

147°

179.5°

opposite angles

angles within a quadrilateral that do not share any common sides

*Vicky explained to Malcolm that she would never ever date him because the two of them are like **opposite angles**— without a common side.*

In the square below, angle *A* and angle *C* are opposite angles.

order of operations

the correct order to use when performing different operations in an expression; a good way to remember the correct order is PEMDAS: *Parentheses* first, then *Exponents,* then *Multiplication* and *Division* (left to right), and lastly *Addition* and *Subtraction* (left to right). If you have difficulty remembering PEMDAS, use this sentence to recall it: *Please Excuse My Dear Aunt Sally*.

*Stephanie has a specific **order of operations** when it comes to doing her homework: first math, then science, then English.*

Simplify the expression $9 - 2 \times (5 - 3)2 + 6 \div 3$.

ordered pair

two numbers (x, y) that when given in a specific order name a location on the coordinate plane; an ordered pair indicates a relationship between the two numbers

*Malcolm knew he'd be great friends with his European IM pen pal when Proust15 gave him the **ordered pair** of (latitude 10°, longitude 50°) rather than saying he was from Germany.*

What ordered pair names the point 3 units to the right and 5 units down from the origin in the coordinate plane?

origin

the point (0, 0) in the coordinate plane where the *x*- and *y*-axes
intersect

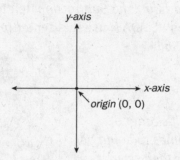

Hannah entered the gigantic Mall of America and quickly noted the
origin *of her shopping spree so that she'd have some hope of finding*
her car in the parking lot that evening.

parabola

the U-shaped graph of a quadratic equation

*Stephanie was thrilled when Diana observed that the punk rocker's U-shaped piercing looked like a cute little **parabola** sticking out of his nose. Finally, the tutoring was showing results!*

Examples of parabolas.

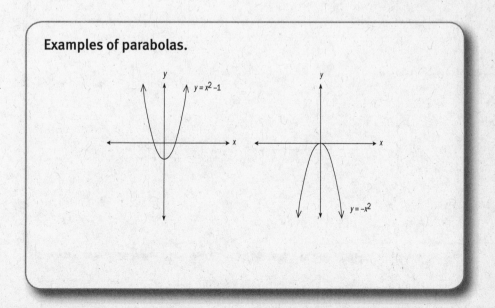

parallel

lines in the same plane that will never intersect

*Brad always thought that gymnastics was easy until he tried to perform a trick on the **parallel** bars and ended up with a broken nose.*

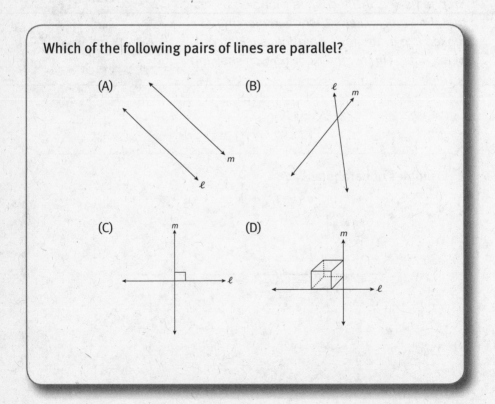

Which of the following pairs of lines are parallel?

(A)

m

ℓ

(B)

ℓ *m*

(C)

m

ℓ

(D)

m

ℓ

parallelogram

a quadrilateral with two pairs of parallel sides; opposite sides are equal, opposite angles are equal, and consecutive angles add up to 180 degrees

The perimeter of a parallelogram is equal to the sum of the lengths of the four sides, which is equivalent to 2(length + width). The area of parallelogram = base x height.

*The girls on the synchronized swimming team formed two pairs of parallel sides in the pool—the perfect **parallelogram**—as an ode to Mrs. Schwartz, their favorite geometry teacher.*

What is the perimeter of a parallelogram with a length of 6 units and a width of 4 units?

part

a portion or piece of a whole

*Brad was pumped after the secret midnight joyride in his dad's Jaguar...until he noticed the three huge gashes he'd made on the back **part** of the car.*

In a class of 24 students, 12 students have brown hair. What part of the class does not have brown hair?

pattern

a predictable sequence

*Vicky and her friends concluded that the guys at their school usually exhibit the same **pattern** of breakup: (1) He stops showing up at your locker. (2) He "forgets" you have plans after school. (3) He says "I'll call you" and then never does again.*

Examples of patterns:

The sequence 0, 2, 4, 6, 8, 10,... is a pattern of even numbers.

The sequence 5, 10, 15, 20, 25,... is a pattern of multiples of 5.

The sequence 1, 3, 9, 27, 81,... is a pattern of the powers of 3.

---

percent

the ratio of a number compared to 100; denoted with the symbol (%).

*Miraculously, Diana guessed "C" for 70 **percent** of the questions on her SAT exam and still came out with an awesome grade...and then she woke up.*

What is 60% of 30?

perimeter

the distance around a figure or object

*Brad contemplated being a lifeguard as he sat in one of the chairs situated around the **perimeter** of the pool. The job would be perfect...if only he didn't have to get his hair wet.*

What is the perimeter of the figure below?

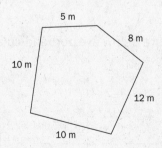

permutations

the possible arrangements of a set of objects, where the order of the arrangement matters; the formula is $_nP_r = \dfrac{n!}{(n-r)!}$ where n is the total number of objects and r is the number of objects to be arranged or ordered

*In order to calm her nerves, Stephanie figured out the various **permutations** of people for the first, second, and third place slots as she waited for the principal to call the winners of the talent contest.*

If there are seven students in an art contest, how many different orders are there for first, second, and third place?

perpendicular

lines or segments that intersect to form right angles

*Brad made sure that he sat **perpendicular** to the pretty girl on the bus so that when she looked up from her book she'd get an awesome view of his profile.*

Lines *l* and *m* in the figure below are perpendicular.

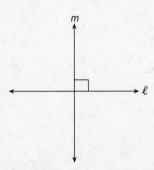

perpendicular bisector

a line or segment at a 90 degree angle that crosses a given segment at its midpoint

*Diana finally understood what a **perpendicular bisector** looked like when her teacher grabbed the two cherry licorice sticks on her desk and formed a plus sign with them.*

If line $\overleftrightarrow{AB}$ is the perpendicular bisector of line segment $\overline{JL}$, and the measure of $\overline{JL}$ is 14 cm, what is the measure of segment $\overline{KL}$?

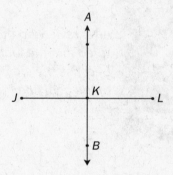

pi

the ratio of the circumference of a circle to its diameter; represented by the symbol π and often approximated as 3.14 or $\frac{22}{7}$

Hannah could not believe the judge gave her gymnastics routine a 3.14—a score equal to the value of pi.

What is the value of the circumference of a circle if the diameter is 10 *m*?

plane

an infinite set of points contained on a flat surface that continues in all directions; any three points not contained in the same line define a plane

After reviewing the play, the referee ruled that the touchdown counted because the football crossed the plane of the goal line.

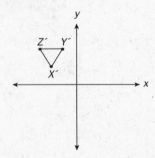

The plane in the figure above can be named plane *XYZ*.

plot

to find the location of a point on a graph or coordinate grid

Wishing that he'd never heard the words "Big Gulp," Brad frantically **plotted** *the next rest stop on his map and prayed he'd reach it before wetting his pants.*

Plot the point *A* (4, −5) on a coordinate grid.

point

a location in space

*Malcolm knew that his brown Honda Civic wasn't cool, but at least it got him from **point** A to point B.*

In the figure below, what point is located at the intersection of lines *l* and *m*?

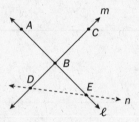

point-slope form

a form of a linear equation that uses a point on the line and the slope of the line; the form is

$y - y_1 = m(x - x_1)$ where (x_1, y_1) is a point on the line and *m* is the slope

*Using the **point-slope form** made it easy for Vicky to quickly identify one point on the line and the line's slope.*

Which of the following represents the point-slope form of the equation of the line with slope of 2 and through the point (4, 3)?

(A) $y - 3 = -2(x - 4)$

(B) $y + 3 = 2(x + 4)$

(C) $y - 3 = 2(x - 4)$

(D) $y + 3 = 2(x + 4)$

polygon

a closed figure whose sides are line segments

*Stephanie had a feeling that the girls huddled in a tight **polygon** were whispering about her.*

Which of the following is not a polygon?

(A) circle

(B) square

(C) pentagon

(D) hexagon

polynomial

an expression containing a monomial or the sum of two or more monomials

*Malcolm was tired of feeling like a monomial in a sea of **polynomials**. He vowed to have a date for the next dance.*

Examples of polynomials:

$2x$, $3y^2 + 1$, $6x^3y^5$, $x - 6$, and $10 - 3z + 4x^3$

positive

a value that is greater than 0, or located to the right of 0 on a number line

*Brad wished the number of girls who accepted his invitation to the dance was **positive** rather than zero.*

Examples of positive numbers:

2, 3.4, $\sqrt{2}$, and $6{,}123$

prime factorization

the process of finding all the prime number factors of an integer

*Diana attempted the process of **prime factorization,** not to find the prime number factors of an integer, but to find the prime qualities of her boyfriend. Unfortunately, there weren't too many.*

Find the prime factorization of 36.

prime number

a whole number whose only factors are 1 and itself

*Malcolm was totally psyched for his 17th birthday because he'd be in the prime once again—the **prime number,** that is.*

Examples of prime numbers:

2, 3, 5, 7, 11, 13, and 17

Note that 2 is the smallest prime number and the only even prime number.

probability

the ratio of the number of ways an event can occur to the total number of outcomes; the formula for the probability of an event E is

$$P(E) = \frac{\text{number of desired outcomes}}{\text{total number of possible outcomes}}$$

Hannah was excited that only 10 students entered the raffle because it meant the **probability** *of her winning was 1 in 10.*

What is the probability of getting a prime number when rolling one die?

product

the result of multiplication

Needless to say, Malcolm's classmates weren't happy to find that he'd accidentally left three deviled eggs for each of his five friends—a **product** *of 15 eggs—in his locker over the weekend.*

What is the product of 52 and 4?

proper fraction

a fraction with a numerator that is less than its denominator

*Vicky didn't get why she was kicked out of the etiquette club for comparing another member to a **proper** fraction. After all, his bottom half is larger than his top half!*

> Examples of proper fractions:
>
> $\frac{1}{3}$, $\frac{10}{17}$, and $\frac{7}{8}$

proportion

an equation stating that two ratios are equal to each other

*Stephanie asked her best friend to give it to her straight: were her eyes in **proportion** or was the left one grotesquely bigger?*

Solve for x: $\frac{5}{6} = \frac{x}{42}$

Pythagorean theorem

The relationship that states the sum of the square of the legs of a right triangle is equal to the square of the hypotenuse; formula is expressed as $(\text{leg}_1)^2 + (\text{leg}_2)^2 = (\text{hypotenuse})^2$ or $a^2 + b^2 = c^2$, where a and b are the legs and c is the hypotenuse

*Malcom tried to impress his date by saying that his uncle Pythagorus invented the **Pythagorean theorem.***

What is the length of the hypotenuse of a right triangle whose legs measure 2 units and 3 units?

quadrant

one of the four sections created by the intersection of the x- and y-axes; each section is labeled from 1 to 4 in roman numerals in counterclockwise order

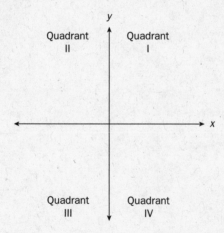

As long as the four kids each grooved in their own **quadrant** *of the tiny dance floor, nobody would get hurt.*

In what quadrant does the point B(−3, 4) lie?

quadratic

an expression or equation of degree 2; contains an x^2 term as the term with the largest exponent

*The Saunders twins considered themselves **quadratic** beauties because they're like the same person squared.*

Which of the following is an example of a quadratic equation?

(A) $y = 2x + 3$

(B) $y = x$

(C) $y = x^3 + 2$

(D) $y = x^2 + 4$

quadratic equation

an equation of degree 2; the standard form is $y = ax^2 + bx + c$

One might say that a double date with the Saunders twins is like going out on a **quadratic equation.**

Each of the following are quadratic equations except:

(A) $y = -3x^2$

(B) $y = 9 - x^2$

(C) $y = 2x + 3$

(D) $y = x^2 + 4x + 4$

quadrilateral

a four-sided polygon; the sum of the interior angles is 360 degrees

*Brad called his architecture teacher closed-minded—not because he got an F in the class, but because she insisted that all buildings must be four-sided **quadrilaterals**.*

Which of the following is not classified as a quadrilateral?

(A) square
(B) trapezoid
(C) pentagon
(D) rhombus

quotient

the answer, or result, of a division problem

*Stephanie divided her 10 new pots of pink lip gloss by 2, then gave the **quotient**, half of the original amount, to her best friend.*

In the statement $45 \div 9 = 5$, what is the quotient?

radical

the symbol $(\sqrt{\ })$ used to find a root of a number, most often the square root

*Malcolm thought it was funny when he told his math teacher that finding the square root of 2,300,056 was a little...**radical**.*

The expression $\sqrt{25}$ means "radical 25," or "the square root of 25," which is equal to 5.

radical equation

an equation that contains at least one radical expression

*Brad admitted that, to him, girls are like **radical equations**. At first they seem easy to understand; then they get mad at you for no reason and you realize that they're tougher to figure out than you'd thought.*

Solve the radical equation $2 + \sqrt{x} = 5$ for x.

radicand

the number or expression contained under the radical sign

*Standing under the store's awning, Stephanie felt a bit like a **radicand** as she waited for the rain to end.*

In the expression $2x + \sqrt{6} - 5^3$, which number is the radicand?

(A) 2

(B) 6

(C) 5

(D) 3

radius

the line segment between the center of a circle and any point on the circle; $\frac{1}{2}$ the length of the diameter of the circle

*Brad took the time to figure out the **radius** of his baby blues so that he could impress the class valedictorian on their date.*

What is the length of the radius of a circle with a diameter of 9 m?

raised to

a phrase used to indicate that a base expression has an exponent

*Vicky knew that 10 measly cupcakes would not make a successful bake sale. So she **raised** each amount in the recipe **to** the third power, baking 1,000 cupcakes in all.*

What is the value of the expression "3 raised to the fourth power"?

random

a sample taken in which each part of a set or grouping has an equal chance of being selected

*Hannah selected a few **random** pieces from her wardrobe, threw them into a bag, and headed out to the sleepover. Too bad she forgot a toothbrush!*

Example:

John was taking a survey. Instead of simply asking his friends to answer the questions, he asked every tenth person who came into the mall that day. Therefore, John was taking a random sample of people to answer the questions.

range

the set of outputs or results of a function

*Vicky put 30 bucks of baby-sitting money into her piggybank. That meant she could spend anywhere in the **range** from $1 through $30 at the mall.*

What is the range of the function $f(x) = x^2$?

rate

a ratio that compares different units; often uses the key word "per"

*Brad wolfed down the chocolate milk at a **rate** of 20 cartons per minute, setting the school record.*

Examples of common rates:

Miles per hour, feet per second, and miles per gallon

ratio

a comparison of two or more numbers or expressions; the ratio comparing a and b can be expressed as $a{:}b$, a to b, or $\frac{a}{b}$

*Stephanie thought she had a pretty good chance of winning a car in the sweepstakes at the mall, until she realized the odds had a **ratio** of 13,004,476,982 to 1.*

Which of the following ratios is equivalent to the ratio comparing 20 to 12?

(A) $\frac{5}{3}$

(B) $\frac{3}{5}$

(C) $\frac{5}{8}$

(D) $\frac{9}{6}$

rational number

a number that can be expressed as a ratio of two integers or as a terminating or repeating decimal

Diana wanted to know if her age, $17\frac{1}{2}$, *is a **rational number**. Does that mean when she becomes 18 she'll be considered irrational?*

> **Examples of rational numbers:**
>
> $-87, \frac{1}{2}$, 4.33333 $(4\frac{1}{3})$, 59, and 100.5 $(100\frac{1}{2})$ are all

rationalize

to convert the denominator of a fraction into a rational number

*Stephanie **rationalized** the fraction so she wouldn't have to work with pesky square roots when figuring out the class budget.*

Rationalize the fraction $\frac{2}{\sqrt{3}}$.

ray

an infinite set of points starting at an endpoint and continuing in a straight path in one direction; ray *AB* is written as $\overrightarrow{AB}$.

*Stephanie slathered on the SPF 50 every day because she didn't want the sun's harmful **rays** to turn her face into a wrinkly mess.*

Which of the following could not be used to name the ray in the figure below?

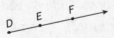

(A) $\overrightarrow{DE}$

(B) $\overrightarrow{FD}$

(C) $\overrightarrow{DF}$

real numbers

the set containing all rational and irrational numbers

Malcolm realized that his feelings for Vicky were as infinite as a set of real numbers.

The rational numbers consist of all repeating and terminating decimals, including integers, whole numbers, and natural numbers. The irrational numbers consist of all nonrepeating, nonterminating decimals such as π, $\sqrt{2}$, 0.1011011101111...

reciprocal

a fraction with the numerator and denominator switched; the product of a number and its reciprocal is 1

Vicky's big sister, Nicky, was thrilled when she got a diamond engagement ring that was 3 carats and relieved because it wasn't its **reciprocal,** $\frac{1}{3}$ *carat.*

Example:

The reciprocal of $\frac{3}{7}$ is $\frac{7}{3}$. The reciprocal of 5 is $\frac{1}{5}$.

rectangle

a parallelogram with four right angles; opposite sides are equal, and the diagonals are equal and bisect each other

The perimeter of a rectangle is equal to the sum of the lengths of the four sides, which is equivalent to 2(length + width).

The area of a rectangle = length × width.

*Hannah loved the pink throw pillows on her bed because they were round and sparkly, unlike the boring white **rectangular** pillows she hid underneath her bedspread.*

What is the area of a rectangle whose length is 7 units and whose width is 3 units?

reflection

a mirror image of an object created by flipping the object over a line or through a point; for example, the symbol $r_{x\text{-axis}}$ denotes a reflection over the x-axis and $r_{y\text{-axis}}$ denotes a reflection over the y-axis

*Brad thought his date with the valedictorian was going great...until she caught him staring at his **reflection** in a soup spoon.*

Which of the following shows the reflection of triangle ABC over the *y*-axis?

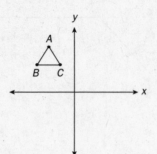

(A)   (B)

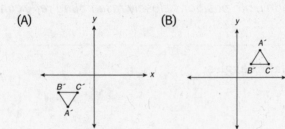

(C)   (D)

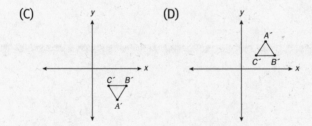

reflex angle

an angle whose measure is more than 180 degrees

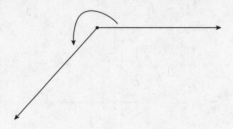

*Hannah did not realize that taking an advanced yoga class meant contorting her body into positions closely resembling **reflex angles**.*

Examples of reflex angles:

191°
214°
270°
359.5°

relatively prime

two or more integers that have no common factor other than 1

*When the valedictorian told Brad that they were like two **relatively prime** integers, she was really trying to say that they didn't have much in common.*

Determine if the integers 35 and 54 are relatively prime.

remainder

the whole number left over after division

*Vicky swore that she didn't eat the **remainder** of the cupcakes herself after the bake sale was over.*

Find the remainder when 183 is divided by 24.

rhombus

a parallelogram with four equal sides

*Malcolm complimented Stephanie on the fancy **rhombuses** on the back of her jeans—in other words, the square pockets.*

Which of the following cannot be classified as a rhombus?

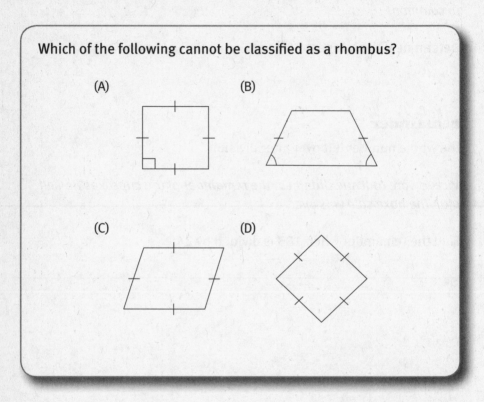

(A)

(B)

(C)

(D)

right angle

an angle that measures 90 degrees

Malcolm never opened his locker wider than a ***right angle****—any more and he'd be practically inviting someone to shove him inside it.*

Which of the following is a right angle?

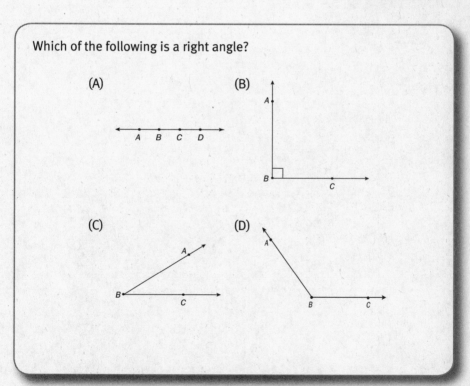

(A)

(B)

(C)

(D)

rotation

an image created by turning the object about a fixed point; for example, the symbol $rot_{180°}$ denotes a rotation of 180 degrees about the origin; a positive degree of rotation is a counter-clockwise turn

*It's a well-kept secret that Vicky can **rotate** a basketball on her index finger for a whole minute.*

What is the image of a block letter Z after a rotation of 90 degrees?

sample space

the list or set of possible outcomes

*Brad quickly calculated the **sample space** for tossing two dice and realized that his chances of rolling "snake eyes" were pretty slim.*

What is the sample space for rolling a fair die?

scale

a ratio used when measuring that compares two different sizes of an object

*Brad was psyched that Miami, his spring break destination, was only 10 inches from his starting point on the map. Then he checked the **scale** and realized that every inch on the map equaled 100 miles on the road, which meant that Miami was 1,000 miles away!*

If the scale of a model airplane is 1:15 and the length of the model is 3 feet, then what is the length of the actual plane?

scalene

a triangle with no congruent, or equal, sides

*The new girls on the synchronized swimming team thought they performed an awesome spinning isosceles triangle in the pool, but the judge said it was a **scalene** because the sides were totally unequal.*

Which of the following measures could represent the sides of a scalene triangle?

(A) 3, 3, 3
(B) 3, 4, 5
(C) 2, 2, 3
(D) 5, 6.1, 6.1

scientific notation

a way to write numbers as the product of a power of 10 and a number greater than or equal to 1 and less than 10

*Instead of writing out tons of zeros to make the class paper's estimated circulation seem high, Malcolm wrote it in **scientific notation** so it would fit one sheet of paper.*

Examples of scientific notation:

2.1×10^4 and 6.78×10^{-5}

sector

a piece of the area of a circle; if *n* is the degree measure of the sector's central angle, then the formula for the area of the sector is $\frac{n}{360}\left(\pi r^2\right)$

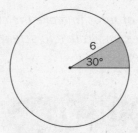

*Brad was so hungry that he ate not only his **sector** of the apple pie but his girlfriend's, too.*

In the figure above, the radius is 6 and the measure of the sector's central angle is 30 degrees. The sector has $\frac{30}{360}$ or $\frac{1}{12}$ of the area of the circle:

$$(\frac{30}{360})(\pi)(6^2) = (\frac{1}{12})(36\pi) = 3\pi.$$

set

a grouping with a common definition

*Stephanie sneaked her new iPod into school and listened to her favorite **set** of dance songs instead of her history lecture.*

Which of the following represents the set of odd numbers?

(A) {0, 1, 2, 3,…}

(B) {0, 2, 4, 6, 8,…}

(C) {1, 1, 2, 3, 5,…}

(D) {1, 3, 5, 7,…}

similar triangles

triangles that have the same shape, corresponding angles that are
equal, and corresponding sides that are in proportion

*Hannah assured her sister that the bracelets
they received for Christmas were equal by
showing her they were **similar triangles** with
the same angles all around.*

The triangles below are similar because they have the same angles.
The 3 corresponds to the 4 and the 6 corresponds to the s: $\frac{3}{4} = \frac{6}{s}$, so
$3s = 24$, and $s = 8$.

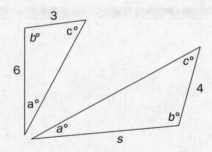

simple interest

the interest calculated by multiplying the principal times the interest rate times the time in years; formula is $I = p \times r \times t$

*The 10 cents of **simple interest** collected in Diana's savings account at the end of the year was disappointing. But what could she expect when she deposited only 10 dollars and earned only 1 percent interest?*

If Shari deposited $500 in her account that earns 6% interest per year, what is the amount of interest earned after 10 years?

simplify

to perform operations in an expression to make a problem less complicated

*Brad knew he had to **simplify** his life somehow, so he decided to break up with two of the three girls he was dating.*

Simplify the expression $3(x + 4) - 2x$

sine (sin)

the ratio of the side opposite to a given angle to the hypotenuse in a right triangle

Vicky thought it was strange when her date used the sine function to announce the angle that the anchor line made with the volleyball net. Was this supposed to impress her?

What is the sine of angle *B* in the following right triangle?

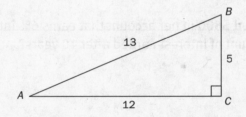

skew lines

lines that are not contained in the same plane that will never intersect; skew lines exist in three dimensions

*Malcolm was starting to think that he and Vicky were kind of like **skew lines**. After all, they existed in totally separate social planes and would never cross paths...that is, if he stopped taking the long way to class.*

In the figure below, the lines *l* and *m* are skew lines.

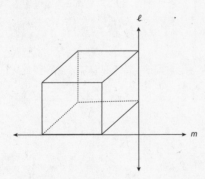

slope

the steepness of a line; the ratio of the change in *y*-values to the change in *x*-values and is found by using the formula

$$m = \frac{\text{change in } y}{\text{change in } x} = \frac{y_1 - y_2}{x_1 - x_2}$$

At the ski resort, Hannah realized that the mountain's **slope** *was much steeper than she'd thought.*

What is the slope of the line between the points (6, 3) and (1, 2)?

slope intercept form

a form of a linear equation that uses the slope and y-intercept of the line; the form is $y = mx + b$ where m is the slope of the line and b is the y-intercept

*Using the **slope intercept form** made it easy for Brad to project how steep the football team's fundraising profits would rise from year to year.*

Which of the following equations is not in slope-intercept form?

(A) $y = x$

(B) $y = 3x - 1$

(C) $y - 3 = 4x$

(D) $y = \dfrac{1}{2}x + 1$

solve

to find the answer to a problem, or find an unknown

*Vicky knew it was useless to try to fix her bad hair day when it was raining. The only way to **solve** the problem was to wear a paper bag over her head.*

Solve for x: 3x − 4 = 8

square

a parallelogram with four equal sides and four right angles; the perimeter of a square is equal to four times the length of one side; the area of a square is equal to one side squared

*Stephanie laughed when her mom told her that **"square"** was slang for a boring person back in the 1960s—because the four equal sides of a square are kind of dull.*

Given the statements below, which of the following is always true?

      I. All parallelograms are squares
      II. All squares are parallelograms
      III. Some parallelograms are squares

(A) I only

(B) I and II

(C) II and III

(D) I and III

squared

a number or expression that is multiplied by itself; an expression raised to the power of 2

*A few of the geeks in the math lab liked to call the Saunders twins "Saunders **squared**."*

What is the value of 6 squared?

square root

a number that, when multiplied by itself, results in the radicand

*Diana had trouble finding **square roots,** until Stephanie explained that they were numbers multiplied by themselves to form a radicand—not the little dangly things hanging off the bottom of a plant.*

Examples:

$\sqrt{9}$ means "the square root of 9" and is equal to 3 since 3 × 3 = 9.

$\sqrt{64}$ means "the square root of 64" and is equal to 8 since 8 × 8 = 64.

$\sqrt{144}$ means "the square root of 144" and is equal to 12 since 12 × 12 = 144.

straight angle

an angle that measures 180 degrees

*Malcolm told his little brother to make like a **straight angle** and go to sleep.*

In the figure below, which angle is a straight angle?

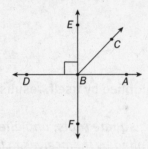

substitution

the process of replacing unknown quantities, or variables, with assigned values

*Brad thought he was so funny when he ordered a hamburger and fries, but made a **substitution** of chicken nuggets for the burger and a salad for the fries.*

Substitute $a = 3$ to find the value of $2a - 1$.

subtract

to take away from one quantity, or the difference between two quantities

*As soon as she was out of her mom's sight, Hannah **subtracted** the big bulky sweater from her outfit to reveal her slinky top underneath. Now she was ready to go to the party!*

Subtract 42 from 98.

sum

the result of addition

*The **sum** of the change at the bottom of Stephanie's purse was not enough to buy a movie ticket so Hannah bought it for her.*

Find the sum of 6 and 17.

supplementary

two angles whose measures sum to 180 degrees

*Malcolm forgot all about his crush on Vicky when he swung open his locker 90 degrees, and the pretty girl next to him did the same then said, "Hey, we just made **supplementary** angles!"*

Two angles that measure 20 degrees and 160 degrees are supplementary since 20 + 160 = 180.

surface area

the sum of the areas of the faces or surfaces of a three-dimensional figure, expressed in square units

*Brad did so many crunches that the **surface area** of his stomach was as hard as a rock.*

If a rectangular prism has a length of 4 *m*, a height of 5 *m*, and a width of 7 *m*, what is the surface area of the prism?

symmetrical

the same on both sides of a line or about a point

*Vicky was a true friend when she admitted that Hannah's ugly blue eyeliner was not **symmetrical**. The line under her left eye was way thicker than the one under her right.*

Examples:

Each of the shapes below show line symmetry.

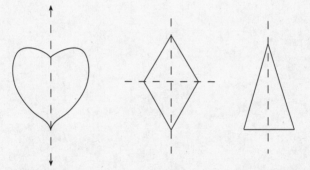

tangent

a line that touches the circle at one point; the radius of the circle is perpendicular to the tangent at the point of the contact; tangent (tan) is also the ratio of the side opposite to the side adjacent in a right angle

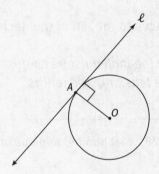

In the figure above, line *l* is tangent to circle *O* at point *A*. Radius $\overline{AO}$ is perpendicular to line *l*.

*Vicky's hand slipped while applying her eyeliner, resulting in the unattractive **tangent** emerging from the outer corner of her eye all the way up to the middle of her forehead.*

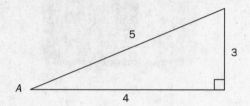

Using the triangle above, what is the value of the tangent of angle *A* (tan *A*)?

tens

the place value two spaces to the left of the decimal point

*Brad thought the girl at the mall must be confused because his looks were clearly rated in the **tens**...not the tenths.*

In the number 7154.628, what number appears in the tens place?

(A) 5

(B) 4

(C) 6

(D) 2

tenths

the place value one space to the right of the decimal point

*Vicky was so angry when the rival team won the cheerleading competition because, according to her, they have three **tenths** of the talent that her team has.*

> **Examples:**
>
> In the numbers 0.345, 6.34, and 141.3, the digit 3 is in the tenths place.

times

the symbol for multiplication (× or •), or the act of multiplying

*The captain of the rival tennis team told Vicky that her team had 10 **times** the talent of Vicky's.*

What is 10 times 8?

transformation

a change in size, position, or orientation made to a geometric object; common transformations are reflections, rotations, translations, and dilations

*Stephanie's **transformation** from short and cute to supermodel beautiful happened when she grew six inches over the summer.*

Which of the following is not classified as a transformation?

(A) translation

(B) dilation

(C) computation

(D) rotation

---

translation

an image created by sliding the object a certain distance in a certain direction; for example, the symbol $T_{3,-2}$ denotes a translation 3 units to the right and 2 units down from the original figure

*While playing chess, Malcolm moved his black knight, resulting in a **translation** of two spaces forward and one to the right from where he began.*

After a translation of $T_{(-5, 4)}$, what is the image of the point (2, 8)?

transversal

a line that intersects, or crosses, two or more other lines

*Stephanie considered herself a **transversal** student because she crossed paths with the jocks, the brains, the theater crowd, and the outcasts.*

In the figure below, line *l* is the transversal that intersects lines *m* and *n*.

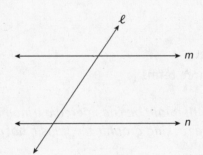

triangle

a three-sided polygon; the sum of the interior angles is 180 degrees

*Hannah couldn't believe that a small **triangle**-shaped patch of grass in New York City was called a park.*

If two interior angles of a triangle measure 35 and 95 degrees, what is the measure of the third angle?

trinomial

a polynomial with three terms

*To some, Diana's outfit might be considered a **trinomial**: one cute little denim mini, plus one rocking graphic tee, plus a hot pair of cowboy boots.*

Which of the following is not an example of a trinomial?

(A) $2x^2 + 3x - 3$

(B) $x^3 - 6x + 2$

(C) $x^3 - 3x$

(D) $4x^6 + 2x^2 - 5x$

union

the set of elements or members of the sets that are in either or both Set A and Set B; can be expressed as A ∪ B

*Brad, being a top student who is great at sports, told the new girl at school that he's the perfect **union** of brains and brawn.*

Example:

If Set A = {1, 2} and Set B = {3, 4}, then A ∪ B = {1, 2, 3, 4}.

unknown

a quantity represented by a variable

*The whole class laughed when the French teacher sat on a whoopee cushion placed on his chair, but the prankster was **unknown**.*

In the expression 2*x*, *x* is the variable, or the unknown. The value of *x* will not be known until a quantity is substituted for *x*, or the expression is set equal to another expression to form an equation where *x* can be solved for.

variable

a letter used to represent a number, or set of numbers, in an equation or expression

*Vicky tried to explain to her mom that good grades alone would not get her into her college of choice. There were other **variables** considered, like her SAT score, extracurricular activities, and teacher recommendations.*

YOUR CAREER OUTLOOK

Examples:

In the expression $6x^2y^3$, x and y are the variables.

In the equation $2z - 4 = 10$, z is the variable.

vertex

the common endpoint of two rays or two segments; also known as the turning point (maximum or minimum) of a quadratic graph (parabola)

*Stephanie planned to "accidentally" bump into the new foreign exchange student at the **vertex** of their paths to second period.*

In the diagram of angle ABC, B is the vertex of the angle.

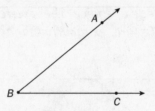

In the diagram of the parabola $y = x^2 + 1$, the point (0, 1) is the vertex (minimum) of the parabola.

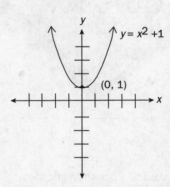

vertical angles

two nonadjacent angles formed by two intersecting lines

*Stephanie quickly grabbed the two slices of pizza that were **vertical angles** to each other, not only because their tips were touching, but because they were the cheesiest.*

Example:

In the figure below, angles 1 and 3 are vertical angles and angles 2 and 4 are vertical angles.

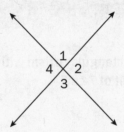

volume

the number of cubic units needed to fill a three-dimensional figure; in a rectangular prism volume is found by multiplying the length × width × height

When it comes to hair, Vicky thinks there's no such thing as too much **volume**–*the more space her curls take up, the better.*

What is the volume of a rectangular prism with a length of 5 cm, a width of 4 cm, and a height of 7 cm?

whole

the total of all parts of a group or quantity

*Brad's date stared in horror as he leaned back in his chair and belched after eating the **whole** plate of pasta.*

In a chorus, 15 members are female and 20 are male. What is the total number of people in the chorus?

x-intercept

the location on a coordinate grid where the graph of an equation crosses the x-axis; in an equation, the value of *x* when *y* = 0

*It was easy for Vicky to center the other cheerleaders along the sideline. All she had to do was stand on an **x-intercept** of the football field, then place six girls on her left and six girls on her right.*

What is the *x*-intercept of the line $y = 3x - 6$?

y-intercept

the location on a grid where the graph of an equation crosses the y-axis; in an equation, the value of y when $x = 0$

From the top of the bleachers Malcolm thought the marching band looked like a vertical line of blips, as if each member was a unique **y**-intercept.

What is the *y*-intercept of the graph of the equation $y - 3 = 2x$?

45-45-90 triangle

a right triangle with three interior angles that measure 45, 45, and 90 degrees and with two congruent sides (legs); the ratio of the sides is $a:a:a\sqrt{2}$

Vicky was disappointed with her leg extensions in aerobics class. If she drew an imaginary line from her right toe to her left, she'd form only a 45-45-90 triangle.

If the measure of the hypotenuse, or longest side, of a 45-45-90 right triangle is $6\sqrt{2}$ units, what is the length of a leg of the triangle?

30-60-90 triangle

a right triangle whose three interior angles measure 30, 60, and 90 degrees; the ratio of the sides is $a : a\sqrt{3} : 2a$

*Malcolm's accordion trio placed their chairs in a **30-60-90 triangle** before beginning their polka for the school talent contest, which looked a little weird as one side was twice as long as the other.*

If the smallest side (leg) of a 30-60-90 right triangle is 5 units, then what is the measure of the other two sides of the triangle?

Answers
and
Explanations

acute angle

(A): 60 degrees is between 0 and 90 degrees. A 90-degree angle is a right angle.

add

34 + 55 = 89

additive inverse

(C): $\frac{1}{2} + -\frac{1}{2} = 0$

adjacent angles

(B)

alternate interior angles

Angles 2 and 3 are alternate interior angles.

alternate exterior angles

Since angle 1 and angle 8 are alternate exterior angles, and line m is parallel to line n, the measure of angle 8 is also 55 degrees.

area

To find the area of a rectangle, multiply the length times the width.

The area is $6 \times 4 = 24$ m.

associative property

(C): Choice B is an example of the distributive property, choice D is an example of the commutative property, and choice A is not an example of a property of numbers.

average

First, find the sum of the set of numbers:

$10 + 12 + 8 + 14 = 44$

Next, divide the sum by the number of values in the set:

$44 \div 4 = 11$

bisector

Since the angle is bisected, or cut into two equal parts, the measure of angle $\angle DBC$ is

$68 \div 2 = 34$ degrees.

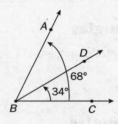

central angle

Since the measure of the intercepted arc is 50°, the measure of angle $\angle ABC$ is also 50°.

chance

Subtract 30% from 100%.

100% − 30% = 70%

There is a 70% chance it will not rain.

circumference

Use the formula

$C = \pi d$

$C = \pi \times 10$

$C = 10\pi$

The circumference is 10π m.

combinations

Using the formula

$$_nC_r = \frac{n!}{r!(n-r)!} =$$

$$_5C_3 = \frac{5!}{3!(5-3)!}$$

$$= \frac{5!}{3!2!}$$

$$= \frac{5 \times 4 \times 3 \times 2 \times 1}{3 \times 2 \times 1 \times (2 \times 1)}$$

$$= \frac{5 \times 4 \times \cancel{3} \times \cancel{2} \times \cancel{1}}{\cancel{3} \times 2 \times 1 \times \cancel{2} \times \cancel{1}}$$

$$= \frac{5 \times 4}{2 \times 1} = \frac{20}{2} = 10 \text{ different ways}$$

commutative property

The number 12 belongs in the answer blank.

$$4 \times (8 + 12) = (8 + 12) \times 4$$
$$4 \times 20 = 20 \times 4$$
$$80 = 80$$

congruent

Since the two triangles are congruent, line up corresponding parts to find the missing measure. Angle X corresponds with angle A, so the $m\angle X = 75°$.

consecutive integers

To solve for the integers, let x = the first integer, let $x + 1$ = the second, and let $x + 2$ = the third.

Since the sum of the integers is 57, then $x + (x + 1) + (x + 2) = 57$.

Combine like terms to get the equation $3x + 3 = 57$.

Subtract 3 from each side of the equation to get $3x = 54$.

Divide each side by 3 to get $x = 18$, so $x + 1 = 19$ and $x + 2 = 20$.

The three integers are 18, 19, and 20.

corresponding

Angles 1 and 5 are corresponding.

cosine

Since the measure of the leg adjacent is 6 and the hypotenuse is 10, then $\cos A = \dfrac{6}{10} = \dfrac{3}{5}$ or 0.60.

cubed

The value of 2 cubed can be expressed as
$2^3 = 2 \times 2 \times 2 = 8.$

cube root

The cube root of 64, or $\sqrt[3]{64}$, is 4, since
$4 \times 4 \times 4 = 64.$

diagonal

The correct answer is 2. In the figure below, this is shown by diagonals $\overline{WY}$ and $\overline{XZ}$.

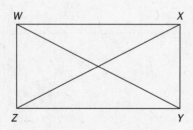

diameter

Since the radius is equal to half of the diameter, then the length of the diameter is $6 \times 2 = 12$ m.

difference

Find the difference between the two numbers by subtracting.

$68 - 33 = 35$

dilation

To find the image of the point, multiply the coordinates by the scale factor of 3.

$(4, -5) \Rightarrow (4 \times 3, -5 \times 3) \Rightarrow (12, -15)$

distance formula

Substitute the values into the formula and evaluate

$$d = \sqrt{(3 - (-1))^2 + (2 - (-1))^2}$$
$$d = \sqrt{(4)^2 + (3)^2}$$
$$d = \sqrt{16 + 9}$$
$$d = \sqrt{25}$$
$$d = 5$$

The distance is 5 units.

distributive property

The number 4 belongs in the answer blank.

$$4(20 + 5) = (4 \times 20) + (4 \times 5)$$
$$4(25) = 80 + 20$$
$$100 = 100$$

divide

$$54 \div 9 = 6$$

dividend

33 is the dividend, or the number being divided into.

divisor

7 is the divisor.

domain

The domain of $f(x) = \dfrac{1}{1 - x^2}$ is all real number values of x except 1 and -1, because for those values the denominator has a value of 0, and is therefore undefined.

edge

The edge is line segment $\overline{DE}$.

equation

In this case, the expressions that are set equal are $3x - 4$ and -16.

To solve an equation, get the variable by itself on one side of the equation.

First, add 4 to each side of the equation.

$$3x - 4 + 4 = -16 + 4$$
$$3x = -12$$

Then, divide each side by 3.

$$\frac{3x}{3} = \frac{-12}{3}$$
$$x = -4$$

equilateral triangle

Since two of the angles measure 60 degrees each, the measure of the third angle must also be 60 degrees. $60 + 60 + 60 = 180$ degrees. Since each of the angles are the same measure, then the triangle has three equal sides. Therefore, side $\overline{YZ}$ measures 5 inches.

equivalent

(D): Note that $\frac{6}{8} = \frac{3}{4}$. The fraction $\frac{16}{20}$ simplifies to $\frac{4}{5}$, while each of the other choices are equal to $\frac{3}{4}$.

exponent

(C): The number 4 is the exponent on the base number of 5.
$5^4 = 5 \times 5 \times 5 \times 5 = 625$.

expression

(A): This choice is an example of an expression. Each of the other answer choices contains an equal sign or inequality symbol.

exterior angle

Since any pentagon has a sum total of 540 degrees in its interior angles, then there are 108 degrees ($540 \div 5$) in one interior angle of a regular pentagon. The measure of the exterior angle is equal to $180 - 108 = 72$ degrees.

face

Each face of a rectangular prism is rectangular in shape. Find the area by multiplying $A = l \times w = 6 \times 10 = 60\text{cm}^2$.

factorial

There are 5 choices for the first spot on the shelf, 4 choices for the second, 3 for the third, 2 for the fourth, and only 1 book left for the fifth. Thus, the total number of arrangements for those five books can be represented by $5! = 5 \times 4 \times 3 \times 2 \times 1 = 120$.

FOIL

First, multiply the first terms $x \times x = x^2$. Then multiply the outer terms to get $x \times -4 = -4x$. Multiply the inner terms to get $5 \times x = 5x$. Then multiply the last terms to get $5 \times -4 = -20$. The expression becomes $x^2 - 4x + 5x - 20$. Combine like terms for a final answer of $x^2 + x - 20$.

function

(C): In this table, an x-value of 1 has a y-value of 3 and another y-value of 9.

greater than

(C): 8 is greater than 7.

greater than or equal to

Subtract 4 from each side of the inequality

$x + 4 - 4 \geq 6 - 4$

$x \geq 2$

Therefore, any number greater than or equal to 2 is a solution to the inequality.

greatest common factor

To find the greatest common factor, break down the integers into their prime factorizations and multiply all the prime factors they have in common.

Therefore: $36 = 2 \times 2 \times 3 \times 3$, and $48 = 2 \times 2 \times 2 \times 2 \times 3$.

What they have in common is two 2s and one 3, so the GCF is $2 \times 2 \times 3 = 12$.

hundreds

The number 6 appears in the hundreds place.

identity

(B): Choice A illustrates the additive inverse property, choice C illustrates the multiplicative inverse property, and choice D represents the multiplicative identity property.

image

(A)

inequality

(D)

inscribed angle

Since the measure of the inscribed angle is equal to half the measure of the intercepted arc, then the angle is equal to $96 \div 2 = 48$ degrees.

in terms of

To do this, isolate x (get x by itself):

$3x - 10y = -5x + 6y$	Add $5x$ and $10y$ to each side
$3x + 5x = 6y + 10y$	Combine like terms
$8x = 16y$	Divide each side by 8
$x = 2y$	

isolate

First, subtract 2x from each side of the equal sign.

$$2x - 2x + 3y = 9 - 2x$$
$$3y = 9 - 2x$$

Then, divide each side by 3.

$$\frac{3y}{3} = \frac{9}{3} - \frac{2x}{3}$$

$$y = 3 - \frac{2x}{3}$$

Therefore, y is isolated and is equal to $3 - \frac{2x}{3}$.

isosceles

(C): Two sides measure 8 in choice C, so the triangle would be isosceles.

least common denominator

To find the least common denominator, find the least common multiple of 5 and 7. Since 35 is the smallest multiple of both 5 and 7, 35 is the least common denominator of the two fractions. Now, convert each fraction to the common denominator in order to perform operations using these fractions.

$\frac{3}{5} = \frac{21}{35}$ and $\frac{5}{7} = \frac{25}{35}$

least common multiple

To find the least common multiple, check out the multiples of the larger integer until one is found that is also a multiple of the smaller. To find the LCM of 12 and 15, begin by taking the multiples of 15: 15, 30, 45, 60, 75,... Fifteen is not divisible by 12; 30 is not; nor is 45. But the next multiple of 15, 60, is divisible by 12, so 60 is the LCM.

leg

False: Use the Pythagorean theorem $a^2 + b^2 = c^2$, where a and b are the legs and c is the hypotenuse. $6^2 + 7^2 = 10^2$

$$36 + 49 = 100$$
$$85 \neq 100$$

less than

Since this inequality is read "ten is less than x," then the only value from the set that works is 11. $10 < 11$. The first term has a smaller value than the second term.

less than or equal to

Divide each side of the inequality by 2.
$x \leq 10$

Therefore, the solution set contains any number less than or equal to 10.

line

The line in the figure could be named line CD, $\overleftrightarrow{CD}$, $\overleftrightarrow{DC}$, or line n.

line segment

True: Although the order does matter when naming certain figures of geometry (rays, for example), the order does not matter as long as the points used are the endpoints.

linear equation

(A)

lowest terms

To reduce a fraction to lowest terms, factor out and cancel the greatest common factor of the numerator and denominator.

Since the GCF of 28 and 36 is 4, then $\dfrac{28}{36} = \dfrac{\cancel{4} \times 7}{\cancel{4} \times 9} = \dfrac{7}{9}$.

mean

First, add the scores: $84 + 90 + 96 = 270$.
Next, divide by 3 since there are 3 test scores. $270 \div 3 = 90$.

median

First list the scores in increasing or decreasing order: 57, 73, 86, 88, 94. The median is the middle number, or 86.

midpoint

The midpoint of (3, 5) and (9, 1) is $\left(\dfrac{3+9}{2}, \dfrac{5+1}{2}\right) = \left(\dfrac{12}{2}, \dfrac{6}{2}\right) = (6, 3)$.

minus

$14 - 8 = 6$

mode

The mode of the scores would be 93 because it appears more often than any other score.

monomial

(B): This choice is not a monomial; there are two terms (x and 4) when a monomial only has one.

multiplicative inverse

Since the multiplicative inverse is the reciprocal of the number, the correct answer is B.

$$-4 \times \frac{-1}{4} = 1$$

multiply

6×3 is equal to 3 groups of 6, or $6 + 6 + 6$ which is 18. Therefore, $6 \times 3 = 18$.

natural numbers

(C): The set of natural numbers does not include zero, negative numbers, fractions, or decimals.

order of operations

Begin with the parentheses:

$(5 - 3) = 2$

Then do the exponent:

$2^2 = 4$

Now the expression is:

$9 - 2 \times 4 + 6 \div 3$

Next do the multiplication and division to get
$9 - 8 + 2$, and finally do the subtraction and addition to get a result of 3.

ordered pair

Keep in mind that positive x-values name points to the right of the origin and negative x-values name points to the left; positive y-values name points above the origin and negative y-values name points below. Thus, the point is (3, −5).

origin

Since the point (−5, 0) is five units to the left of the origin as shown in the figure below, the answer is 5.

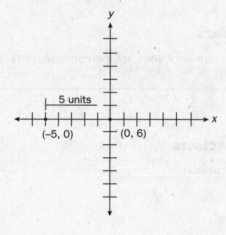

parallel

(A): Choices B and C are examples of lines that intersect, and the lines in choice D are skew lines which do not intersect but are not located in the same plane.

parallelogram

Since the perimeter is 2(length + width), then it is equal to 2(6 + 4) = 2(10) = 20 units.

part

Since there are a total of 24 students, subtract 24 − 12 = 12 to find the number of students that do not have brown hair. Therefore, 12 students or $\frac{1}{2}$ of the class does not have brown hair.

percent

Since 60% is equal to $\frac{60}{100}$, set up the proportion $\frac{60}{100} = \frac{x}{30}$.

Cross-multiply to get 100x = 1800.

Divide each side of the equation by 100 to get $x = 18$.

perimeter

The perimeter is found by finding the sum of the measures of the sides of the figure.

The perimeter of this figure is found by adding
$8 + 12 + 10 + 10 + 5 = 45$ m.

permutations

Using the formula,

$$_nP_r = \frac{n!}{(n-r)!}$$

$$_7P_3 = \frac{7!}{(7-3)!}$$

$$= \frac{7!}{4!}$$

$$= \frac{7 \times 6 \times 5 \times 4 \times 3 \times 2 \times 1}{4 \times 3 \times 2 \times 1}$$

$$= \frac{7 \times 6 \times 5 \times \cancel{4} \times \cancel{3} \times \cancel{2} \times \cancel{1}}{\cancel{4} \times \cancel{3} \times \cancel{2} \times \cancel{1}}$$

$$= 7 \times 6 \times 5 = 210 \text{ different ways.}$$

perpendicular bisector

Since line $\overleftrightarrow{AB}$ is the perpendicular bisector of the segment, then the length of $\overline{KL}$ is equal to $\frac{1}{2}$ of the length of the entire segment. Therefore, the measure of $\overline{KL}$ is 7 cm.

pi

Since pi is equal to the circumference divided by the diameter, use $\pi = \dfrac{C}{d}$ and cross-multiply to get $C = \pi d$.

Since $d = 10$, $C = 10\pi$ or $10 \times 3.14 = 31.4$ units.

plot

Start at the origin (0, 0). Move 4 units to the right and 5 units down to find the location of point A.

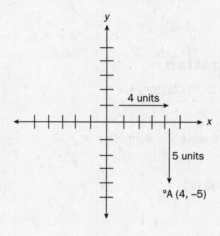

point

Point B is located at the intersection of lines *l* and *m*.

point-slope form

(C)

polygon

(A): A circle does not have sides that are line segments.

prime factorization

Start by breaking it into the factors 4 × 9:

36 = 4 × 9

Then, break down the factors of 4 and 9:

2 × 2 × 3 × 3

The prime factorization is $2^2 \times 3^2$.

probability

Since there are 3 prime numbers (2, 3, and 5) out of a total of 6 sides, then the probability is P(prime number) = $\frac{3}{6}$ or $\frac{1}{2}$.

product

$52 \times 4 = 208$

proportion

The equation $\dfrac{5}{6} = \dfrac{x}{42}$ is a proportion and can be solved for x by cross-multiplying:

$5 \times 42 = 6 \times x$

$210 = 6x$

Divide each side of the equation by 6:

$35 = x$

Pythagorean theorem

If c represents the hypotenuse, and one leg is 2 and the other leg is 3, then:

$2^2 + 3^2 = c^2$

$c^2 = 4 + 9$

$c = \sqrt{13}$

quadrant

To find the location of this point, start at the origin (0,0). Move 3 units to the left and 4 units up. The location of point B is in quadrant II.

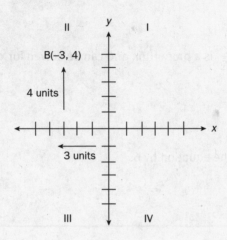

quadratic

(D): Choice D contains an x^2 term as the term of the largest degree.

quadratic equation

(C): It is the only equation without an x^2 term as the term with the highest degree.

quadrilateral

(C): A pentagon has 5 sides.

quotient

5 is the quotient, or the result of division.

radical equation

Solve radical equations by isolating the radical expression and using standard rules of algebra.

First, subtract 2 from each side of the equation:

$2 - 2 + \sqrt{x} = 5 - 2$

The equation becomes $\sqrt{x} = 3$.

Next, square each side to eliminate the radical. $\left(\sqrt{x}\right)^2 = 3^2$.

Therefore, $x = 3^2 = 9$.

radicand

(B): 6 appears under the radical sign.

radius

To find the radius when the diameter is known, multiply $\frac{1}{2}$ times the diameter.

$\frac{1}{2} \times 9$ is 4.5 m. The radius is 4.5 m long.

raised to

This expression translates to 3^4, which is equal to $3 \times 3 \times 3 \times 3 = 81$.

range

The range of $f(x) = x^2$ is all numbers greater than or equal to zero, because x^2 cannot be negative.

ratio

(A): The given ratio of 20 to 12 can also be written as $\frac{20}{12}$, which simplifies to $\frac{5}{3}$.

rationalize

Multiply numerator and denominator by $\sqrt{3}$.

$$\frac{2}{\sqrt{3}} \times \frac{\sqrt{3}}{\sqrt{3}} = \frac{2\sqrt{3}}{\sqrt{9}} = \frac{2\sqrt{3}}{3}$$

The denominator is now a rational number.

ray

(B): Since the endpoint of the ray must be listed first, choice B cannot be used to name the ray.

rectangle

The area of a 7-by-3 rectangle is $7 \times 3 = 21$ square units.

reflection

(D): Choice A is a reflection over the *x*-axis, choice B is a translation, or slide, and choice C is a rotation of 180 degrees.

relatively prime

To determine whether two integers are relatively prime, break them both down to their prime factorizations.

Therefore: $35 = 5 \times 7$, and $54 = 2 \times 3 \times 3 \times 3$.

They have no prime factors in common, so 35 and 54 are relatively prime.

remainder

How many times does 24 go into 183? (7 times)

$7 \times 24 = 168$

How many are left over?

$183 - 168 = 15$

The remainder is 15.

rhombus

(B): A trapezoid does not have both sets of opposite sides parallel nor four equal sides.

right angle

(B): A right angle is formed by perpendicular lines, rays, or segments and is frequently shown with a box drawn in the corner of the angle.

rotation

A full rotation is 360 degrees. A rotation of 90 degrees is a $\frac{1}{4}$ –turn counter-clockwise. The block letter Z would appear as this:

Z $\xrightarrow{\text{Rotate } 90°}$ N

sample space

Since the sample space is the set of possible outcomes, it is {1, 2, 3, 4, 5, 6}.

scale

First, set up the proportion $\dfrac{\text{model length}}{\text{actual length}} = \dfrac{1}{15} = \dfrac{3}{x}$.

Next, cross-multiply to get $x = 45$ ft.

scalene

(B): Each of the sides has a different measure.

set

(D)

simple interest

First, change the interest rate from a percent to a decimal:

6% becomes 0.06.

Next, use the formula $I = p \times r \times t$ and substitute the values:

$I = (500)(0.06)(10)$
$I = 300$

She earns $300 in interest after 10 years.

simplify

First, multiply using the distributive property:
$3x + 12 - 2x$

Next, combine like terms:

$x + 12$

sine

Since the measure of the leg opposite is 12 and the hypotenuse is 13, then

$\sin A = \dfrac{12}{13} \approx 0.92.$

slope

$m = \dfrac{\text{change in } y}{\text{change in } x} = \dfrac{y_1 - y_2}{x_1 - x_2} = \dfrac{3 - 2}{6 - 1} = \dfrac{1}{5}$

The slope of the line is $\dfrac{1}{5}$.

slope intercept form

(C): Choice A is in slope intercept form, even though the value of b (the y-intercept) is 0.

solve

Add 4 to each side of the equation.

$3x - 4 + 4 = 8 + 4$

$\quad\quad 3x = 12$

Divide each side by 3.

$\quad\quad x = 4$

square

(C): Not all parallelograms have all of the properties of squares.

squared

6 squared is equal to 6^2 or 6×6, which is equal to 36.

straight angle

In the figure, angle *ABD* is a straight angle since it measures 180 degrees and is formed by a straight line. Angle *EBF* is also a straight angle.

substitution

First, replace, or substitute, 3 for the letter *a* in the expression:

$2(3) - 1$

Then, multiply $2 \times 3 = 6$ and subtract 1 to find the value:

$6 - 1 = 5.$

subtract

$98 - 42 = 56$

sum

6 + 17 = 23

surface area

Since the figure is a rectangular prism, there are a total of six surfaces, or faces. Find the area of each face and add them together.

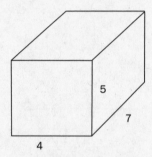

$(4 \times 5) + (5 \times 7) + (4 \times 7) + (4 \times 5) + (5 \times 7) + (4 \times 7)$
$= 20 + 35 + 28 + 20 + 35 + 28 = 166 \text{ m}^2.$

tangent

Tangent is the ratio of the leg opposite the angle to the leg adjacent to the same angle. Since the measure of the leg opposite is 3 and the leg adjacent is 4, then $\tan A = \dfrac{3}{4}$ or 0.75.

tens

(A)

times

$10 \times 8 = 80$

transformation

(C)

translation

To find the image, subtract 5 from the x-value of the point and add 4 to the y-value.

$(2, 8) \Rightarrow (2 - 5, 8 + 4) \Rightarrow (-3, 12)$

triangle

First, find the sum of the two known angles:

$35 + 95 = 130$

Next, subtract the sum from 180:

$180 - 130 = 50$

The third angle measures 50 degrees.

trinomial

(C): Choice C only has two terms, and a trinomial has three.

volume

Use the formula *volume = length × width × height*, or *V = lwh*:

$V = 5 \times 4 \times 7$

$V = 140 \text{ cm}^3$

whole

Add the parts $15 + 20 = 35$ to get the whole, or total amount of people in the chorus.

x-intercept

To find the x-intercept, plug $y = 0$ into the equation and solve for x:

$$y = 3x - 6$$
$$0 = 3x - 6$$
$$0 + 6 = 3x - 6 + 6$$
$$6 = 3x$$
$$2 = x$$

The x-intercept is 2. The graph of this equation will cross the x-axis at the point $(2, 0)$.

y-intercept

To find the y-intercept, you can either put the equation into $y = mx + b$ (slope-intercept) form—in which case b is the y-intercept—or you can just plug $x = 0$ into the equation and solve for y.

In this case, add 3 to each side of the equation to get $y = 2x + 3$. Since the value of b is 3, this graph crosses the y-axis at $(0, 3)$.

45-45-90 triangle

Since the ratio is $a: a: a\sqrt{2}$, and the longest side is $6\sqrt{2}$, then the length of a leg must be 6 units.

30-60-90 triangle

Since the smallest side is 5, the other leg will measure $5 \times \sqrt{3}$, or $5\sqrt{3}$. The largest side, or hypotenuse, will measure twice the length of the smallest side or $2 \times 5 = 10$ units. Therefore, the length of the other two sides are $5\sqrt{3}$ units and 10 units.

NOTES: